YOU CAN SAVE THE Planet

Scholastic Inc.

This book is printed on paper sustainably sourced and certified by the
Forest Stewardship Council.

ISBN 978-1-338-66326-6

10 9 8 7 6 5 4 3 2 1   20 21 22 23 24

Printed in the U.S.A.   40
First printing 2020

Book design by Deena Fleming
Written by Alessandra Potenza

Photos ©: cover background: diversepixel/Shutterstock; cover bottom, 1 bottom: NPeter/Shutterstock;
back cover top left, 2 bottom left: Valmedia/Shutterstock; back cover top right, 2 bottom right: Ink
Drop/Shutterstock; back cover bottom: Bianca Alexis; 1 background and throughout: Art Stocker/
Shutterstock; 3 top background and throughout: mauveineimages/Shutterstock; 4–5 maps: petch
one/Shutterstock; 4 left: Christian Roberts-Olsen/Shutterstock; 4 right: robert cicchetti/Shutterstock;
5 center left: twenty1studio/Shutterstock; 5 center right: lavizzara/Shutterstock; 5 bottom left: Rebekah
Zemansky/Shutterstock; 6 top: Marina Poushkina/Shutterstock; 6 bottom: NASA; 7 top: small smiles/
Shutterstock; 7 center: Ondrej Prosicky/Shutterstock; 7 bottom: Designua/Shutterstock; 8: ARM Photo
Video/Shutterstock; 8 doodles and throughout: antishock/Shutterstock; 9: Marcus Yam/Getty Images;
10: aapsky/Shutterstock; 11: Alessandro Pietri/Shutterstock; 12: Tom Wang/Shutterstock; 13: Ralph
Freso/Getty Images; 14: Doubletree Studio/Shutterstock; 15: Boston Globe/Getty Images; 16: Sean
Gallup/Getty Images; 17: Joe Raedle/Getty Images; 18 top: Monkey Business Images/Shutterstock;
18 bottom: R_Tee/Shutterstock; 19 top: crazystocker/Shutterstock; 19 bottom: Ben Gingell/
Shutterstock; 20–21: Scholastic Inc.; 21 bottom right: petovarga/Shutterstock; 22: NOAA;
23: Chesnot/Getty Images; 24: Johannes Eisele/Getty Images; 25: EuropaNewswire/Gado/Getty
Images; 26 top: Eric Baradat/Getty Images; 26 bottom and throughout: Elenarts/Shutterstock;
27: Robin Loznak/Zuma Press; 28: Rachel Murray/Getty Images; 29: Erik McGregor/Getty Images;
30: Mike Jett/Shutterstock; 31: Mike Jett/Shutterstock; 32: Gemma Chua Tran; 33: Jenny Evans/
Getty Images; 34: Ayakha Melithafa; 35: Sydelle Willow Smith/The New York Times/Redux; 36, 37:
Dinesh Chandra Pandey; 38 center: Dannyphoto80/Dreamstime; 39 top right: Juliasart/Dreamstime;
40 top right: pupahava/Shutterstock; 40 center: Dg Studio/Dreamstime; 40 center right: Eyewave/
Dreamstime; 41 top: Brown Bird Design; 41 bottom: Blueringmedia/Dreamstime; 42: Monkey Business
Images/Shutterstock; 43 top: Ben Gingell/Shutterstock; 43 bottom: Lisa F. Young/Shutterstock;
42–45 background: Mtsaride/Shutterstock; 44 doodle and throughout: mhatzapa/Shutterstock;
47 airplane: ArtMari/Shutterstock.

WHAT'S INSIDE

OUT OF WHACK!

What's going on with planet Earth? It may seem like the past few years have been a near-constant environmental disaster.

In California, terrible **wildfires** have been ravaging thousands of acres of land, filling the air with an eerie, orange smoke. In Puerto Rico, powerful **hurricanes** have destroyed homes and left the entire island without power. Intense **heat waves** in Arizona have made it too hot for kids to play outside, while massive **blizzards** have swept the East Coast. People in certain areas in Florida have suffered from flooded streets and homes.

Disasters seem to be everywhere! That's why kids—just like you!—are jumping into action to save our planet. From Greta Thunberg in Sweden to Jean Hinchliffe in Australia, from the United Nations to their own backyards, kids of all ages and backgrounds across the world are doing their part to make a difference. Read on to learn more and discover how you can help, too!

California Wildfires

Blizzard in New York

But first, what's the difference between climate and weather?

Climate is the usual weather in a region over many years. It influences what types of clothes you have in your closet. For example, if you live in a state with a tropical climate, like Hawaii, it's mostly hot out. You own lots of shorts and T-shirts.

Weather is the day-to-day variation in temperature, wind, and rainfall. Even if you live in a place with a warm climate, the weather could sometimes be rainy and cool.

Hurricane Maria Makes Landfall in Puerto Rico in 2017

Heat Wave in Arizona

WHAT'S UP WITH THE PLANET?

First, let's solve this mystery. Why is extreme weather sweeping the U.S. and other areas of the world?

THE PROBLEM

Scientists have a hunch about what's going on: Earth's average temperature and weather patterns are changing. This is called **climate change**. But how do scientists know climate change is happening?

THE EVIDENCE

Scientists have been using all kinds of instruments to study Earth for a long time. For example, they measure the types of gases that make up Earth's **atmosphere**. Some of these gases, called **greenhouse gases**, trap heat from the sun. Without them, the planet would be a frozen hunk of rock ... and way too cold to live on. But scientists have noticed that the amount of greenhouse gases like carbon dioxide (CO_2) in the air has been increasing for several decades. That's causing Earth's temperature to rise. It's like our planet has a fever!

THE CULPRIT

So why are there extra greenhouse gases in the atmosphere? It's because of us! Humans burn **fossil fuels** like coal, natural gas, and oil to power cars and create electricity. That releases greenhouse gases into the air. It's causing Earth to warm up faster than ever before.

MYSTERY SOLVED!

A warmer world leads to more extreme weather and other problems, such as wildfires and sea levels rising. Why exactly? You're about to find out.

The Greenhouse Effect

Since 1880, Earth's average global temperature has increased by 1.8° Fahrenheit (1° Celsius). Here's how greenhouse gases like CO_2 are causing our planet to heat up.

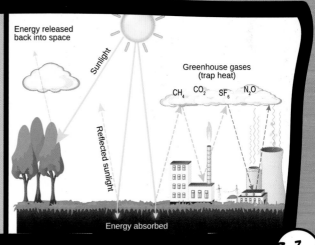

Energy released back into space

Sunlight

Greenhouse gases (trap heat)

CH_4 CO_2 SF_6 N_2O

Reflected sunlight

Energy absorbed

UP IN FLAMES

In November 2018, a wall of flames devoured the town of Paradise, California. Within hours, the raging fires had destroyed almost 14,000 homes—and left many people dead. Wildfires like the one in Paradise are getting bigger and more destructive because of climate change.

It happens like this: Warmer temperatures dry out trees and shrubs, causing them to easily catch fire. **Droughts**, or long periods without rain, can also kill plants, littering the forest floor with kindling ready to burn. These conditions increase the risk of large and dangerous wildfires. Other factors like strong winds help the flames spread far and wide.

Wildfires aren't just a problem in California. Many U.S. states, like Texas and North Carolina, are also struck by wildfires, as are countries like Spain and Australia.

DID YOU KNOW?

About 85 percent of wildfires are caused by humans. Unattended campfires or downed power lines are often to blame.

Firefighters in August 2018 during the Holy Fire near Lake Elsinore, California.

MONSTER STORMS

In recent years, hurricanes barreling through the Atlantic Ocean have been making records. In 2017, for example, Hurricane Maria dumped record amounts of rain over Puerto Rico. The deluge destroyed thousands of homes, triggered landslides and flooding, and helped disrupt drinking water on the island.

Scientists say that these destructive hurricanes are exactly the kind of thing we should expect in a warming world. Why? One reason is that the world's oceans are becoming warmer. Hurricanes form from the moist air that **evaporates** and rises up from warm seawater. Warmer oceans lead to more moisture, which provides more fuel for hurricanes. The end result is stronger hurricanes that pick up more water and dump more rain.

Scientists aren't sure whether a warming world will produce *more* hurricanes. But they do think that the storms will become more intense because of climate change.

DID YOU KNOW?

Scientists who study weather, called **meteorologists**, began giving hurricanes names in the 1950s. That makes communication easier if two or more hurricanes are happening at the same time.

Flooded streets in San Juan, Puerto Rico, after Hurricane Maria devastated the island in November 2017.

FEELING THE HEAT

The past few years have been among the hottest on record. As the planet continues to warm up, that trend is likely to continue.

As part of this warming, heat waves are becoming more common and more dangerous. These long periods of extremely hot weather can destroy crops and harm people's health. Young children, the elderly, and people with certain medical conditions are particularly at risk of heatstroke. This sometimes deadly condition happens when the body can't cool itself down.

Extreme heat can also disrupt transportation. In June 2017, a heat wave in Phoenix, Arizona, caused temperatures to reach up to 119° Fahrenheit (48.3° Celsius). As a result, dozens of flights had to be canceled. Some airplanes can only safely fly at 118° Fahrenheit (47.8° Celsius) or below. Why? Hot air is less dense, or has fewer **molecules** in a given space, than cold air. That's because as air warms up, its molecules move around faster and spread out. So if it's really hot, airplanes need more speed—and longer runways—to produce enough force to lift off the ground.

 DID YOU KNOW?

The number of heat waves in 50 major U.S. cities has tripled since the 1960s from about two per year to six per year.

In June 2017, heat waves hit
Phoenix, Arizona.

GLOBAL WEIRDING

Even if the world is getting warmer, that doesn't mean we're not going to have winters anymore. In fact, climate change increases the risk of extreme snow.

Why? One reason is that the world's oceans are getting warmer. Remember: Warmer water leads to more moisture in the air. Another reason is that rising temperatures cause the air to *hold* more moisture. This means that if a storm forms and it's really cold, the storm will dump that extra moisture as heavy snow.

Scientists also think that there might be a link between unusually warm air in the Arctic and cold weather in the U.S. The warming Arctic might disrupt the way air circulates in the atmosphere. That might cause freezing cold air to rush down south—triggering severe cold weather in areas of the U.S.

Scientists are still researching how a warming Arctic might be linked to cold spells. But one thing is clear: Periods of cold and heavy snow do not mean that global warming isn't happening.

 ## DID YOU KNOW?

Syracuse, New York, is one of the snowiest cities in the U.S. On average, it receives about 124 inches (3 meters) of snow every year. That's about as tall as a basketball hoop!

People dig out their cars after heavy snow in 2018 in Boston, Massachusetts.

RISING WATERS

Around the world, sea levels have risen on average by about 8 inches (20 centimeters) since 1880. That doesn't sound like much. But the rising waters are already making flooding worse in certain areas.

This also has to do with oceans becoming warmer. Warm water expands, so oceans are literally taking up more space. Huge masses of ice, called **glaciers**, are also melting around the world. Some are disappearing right in front of our eyes! The meltwater flows into the ocean and causes sea levels to go up.

More than 600 million people—10 percent of the world's population—live in low-lying coastal areas. Some communities on the South Pacific islands of Fiji are already moving to higher grounds because the oceans are flooding their homes. The same is happening in Louisiana and Alaska.

Sea level rise is causing areas of Miami to flood regularly at high tide, even when it's not raining. That is called "sunny day flooding."

 DID YOU KNOW?

If all the glaciers and ice caps on Earth melted, sea levels would rise about 230 feet (70 meters). That's about as tall as a 23-story building!

People walking through a flooded street in Fort Lauderdale, Florida, in 2015. The flooding is believed to be caused by a combination of rising sea levels and seasonal high tides.

BEAT THE BLUES

If reading about climate change makes you feel overwhelmed, that's normal. This sense of worry about the climate crisis is often called "climate anxiety." "Many people feel anxious about climate change because it is making the future uncertain," says Susan Clayton. She's a professor of psychology, the science of mind and behavior, at the College of Wooster in Wooster, Ohio.

If you're feeling stressed out or sad about climate change, Clayton has a few tips on how to stay upbeat and feel empowered.

1. DON'T KEEP IT INSIDE:

If you're feeling anxious, tell someone in your family or another supportive adult, like a trusted teacher. Discussing your emotions can make you feel better.

2. UNPLUG:

It's easy to get overwhelmed when reading or hearing about the climate crisis. The solution? Take your mind off negative information and do something that makes you happy. Read a book, dance to music, or spend time with your friends.

3. GO OUTSIDE:

When stress becomes too much, there's an easy fix: Take a walk in a forest, in a local park, or on a beach. Nature is a like a free-for-all stress ball! Hearing the birds chirp, listening to the waves lapping the shore, or looking at clouds change into weird shapes can help ease stress. Plus, hiking or strolling is a form of exercise. Being active keeps us healthy!

4. DO SOMETHING ABOUT IT:

One way to feel empowered in the fight against climate change is by taking action. You can decide to change your lifestyle to help protect the environment. Or you might want to join a demonstration with an adult in your family to demand action on climate change. Turn to page 38 to learn about how you can make a difference.

HOW TO SOLVE THE CLIMATE CRISIS

Fortunately, scientists know how we can address climate change. Some of these solutions are already being implemented around the world. Others are still being developed and studied.

Cool Trees: One of the easiest solutions to climate change is planting trees and protecting existing forests from being cut down. Trees absorb carbon dioxide from the air and use it to make their own food in a process called photosynthesis. As a result of that process, trees release oxygen and store carbon in their trunks, leaves, and roots. By removing CO_2 from the atmosphere, trees can help keep our planet cool.

Carbon Vacuums: As emissions of CO_2 keep increasing, engineers are coming up with ways to suck this dangerous greenhouse gas out of the air. The technology, called carbon capture, is already being used around the world. But it's not very efficient—at least not yet. For example, a facility in the European country of Switzerland can only capture about 1,000 tons of CO_2 a year. That's the same amount of CO_2 produced by about 55 people in the U.S. every year.

Heat Shields: Some scientists think that in order to cool the planet, we might have to take drastic action. One idea involves spraying particles up high in the atmosphere to reflect sunlight back into space. That would help bring the world's temperatures down. The process mimics what happens with very big volcanic eruptions. In 1991, Mount Pinatubo, a volcano in the Philippines, spewed so many particles into the atmosphere that it caused global temperatures to drop by about 1° Fahrenheit (.6° Celsius). The idea hasn't been tested on a large scale yet. Some experts fear that it might have unwanted effects.

Clean Energy: Many human emissions of planet-warming greenhouse gases come from the burning of fossil fuels like oil and coal to produce electricity and power cars and airplanes. But there's some good news: We don't have to burn fossil fuels to create energy! We can use **renewable energy** sources instead, like wind and solar power. Electric vehicles are also becoming more popular. And scientists and engineers are trying to develop airplanes powered by the sun.

THE PARIS AGREEMENT

On December 12, 2015, almost two hundred countries reached a historic deal: They agreed to cut greenhouse gas emissions in order to address climate change.

Going Up

Emissions of carbon dioxide have been rising steadily since 1980.

GLOBAL MONTHLY MEAN CO_2

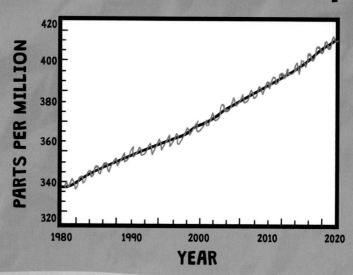

DID YOU KNOW?

The unit to measure how much CO_2 is in the atmosphere is "parts per million." It tells you how many particles of CO_2 are present in one million particles of air.

During the World Climate Change Conference in December 2015 that led to the Paris Agreement, the slogan "For the Planet" was projected onto the Eiffel Tower.

The major goal of the Paris Agreement is to keep the world's temperatures from rising to dangerous levels. Scientists think that if the world warms by 3.6° Fahrenheit (2° Celsius), the consequences—like extreme weather, sea level rise, and shortages of food and water—would be too dire.

However, experts fear that the Paris Agreement isn't enough to stop global warming. Some countries aren't cutting their emissions as much as they said they would. China still relies heavily on power plants that use coal to produce electricity. Those plants produce a lot of CO_2. In 2019, the Australian government approved the opening of a new coal mine. Meanwhile, the U.S. decided to withdraw from the Paris Agreement altogether.

However, more and more local governments, city mayors, and even business leaders in the U.S. and abroad are pledging to cut their own greenhouse gas emissions. There's still a lot to do. But together, we can make change happen!

Greta Thunberg (center) on September 20, 2019, during the Global Climate Strike march in New York City.

MEET THE CLIMATE HEROES

On Friday, September 20, 2019, hundreds of thousands of young people skipped school and took to the streets across the globe. They carried signs that read "There's No Planet B" and "If You Don't Act Like Adults, We Will." An estimated four million people marched together in more than 150 countries on all seven continents—including Antarctica! Their goal: to sound the alarm on climate change and ask global leaders to take action.

The Global Climate Strikes were among the largest protests for climate action in history. But this movement started with just one girl. In August 2018, when she was fifteen years old, Greta Thunberg began skipping school on Fridays. She sat in front of Parliament in her home country of Sweden to demand action on climate change. Her lone protests have inspired thousands of young people to do the same.

In New York City, for example, Alexandria Villaseñor strikes every Friday. She sits in front of the United Nations, an organization of countries that work together on problems like climate change. "Greta gave permission to students all around the world to strike as well," Alexandria says. Here are the stories of other young climate **activists** in the U.S. and abroad—and what they're doing to make change.

Greta Thunberg (left) and Alexandria Villaseñor (right) at a protest at the United Nations Headquarters in New York City in September 2019.

LEVI DRAHEIM

Levi Draheim decided to become a climate activist for a simple reason: He doesn't want his home to disappear. Levi live in Indian Harbour Beach, on an island off the east coast of Florida. Rising seas and stronger storms could one day wash away the island. Levi has already had to evacuate several times because of hurricanes. "It's extremely scary not knowin if you're going to have a home in the future," he says.

In 2015, when he was eight years old, Levi sued the U.S. government with dozens of other young people. The lawsuit argues that the government isn't doing enough to protect its citizens from climate change. It could force lawmakers to pass legislation to cut greenhouse gas emissions.

Levi also goes to demonstrations to demand action on climate change. He's homeschooled, so he doesn't strike every Friday like Greta Thunberg and Alexandria Villaseñor. Instead, he takes time to pick up trash on the beach or on a bridge by his house.

Year born: 2007
From: Indian Harbour Beach, Florida
Favorite subject: Reading comprehension
Favorite outdoor activity: Swimming
Favorite thing about planet Earth: The water springs in Florida
Favorite food: Vegan ramen
Role model: Xiuhtezcatl Martinez (environmental activist)
Dream job: Bird expert
Tip for aspiring climate activists: "Don't ever let somebody tell you that you do not have a voice because that's not true."

QUICK Q&A WITH LEVI

Q. When did you decide to become a climate activist?

A. I decided to become a climate activist when I realized that the adults aren't taking enough action to save our future, our planet, our right to life, liberty, property, and the pursuit of happiness.

Q. What do you do to address climate change in your personal life?

A. My parents and I walk when we can instead of driving. We take short showers and turn off the water when we brush our teeth. We're **vegans**. We also try to eat local. We go to farmers markets and get food that is grown very close.

Q. Do you ever feel overwhelmed by climate change? How do you cope?

A. I feel extremely overwhelmed. The way that I cope is, I go hang out with friends and I spend time in nature. I always have to remember that this is my future, this is my life, and if I'm not going to take action for my life and my future, then who will?

Q. What do you hope the youth climate movement will achieve?

A. We want to show the U.S. government and the people in charge that we care and we're going to take action. We're willing to sacrifice our education to have a future, to have a life.

JAMIE MARGOLIN

In the summer of 2017, the air in Seattle, Washington, was shrouded in haze. Winds were blowing smoke from more than twenty wildfires burning in Canada. The smoke made breathing difficult for Jamie Margolin. But it also encouraged her to take action.

She posted on her Instagram that she wanted to start a youth climate movement. A teen named Nadia Nazar reached out and told her she wanted to help! Together, they started an organization called the Zero Hour. The group helps organize marches and "lobby days," where young people talk to politicians to demand action on climate change.

Why the name the Zero Hour? "We have zero hours left to act on the climate crisis," Jamie says.

Year born: 2001
From: Seattle, Washington
Favorite subjects: English and literature
Favorite outdoor activity: Lying on the grass and cloud watching
Favorite thing about planet Earth: Its diversity and beauty
Favorite food: Her grandma's empanadas
Role model: Alexandria Ocasio-Cortez (U.S. Representative from New York)
Dream job: President of the United States
Tip for aspiring climate activists: "Use your already existing passions and what you're good at to spice up the work that you do."

QUICK Q&A WITH JAMIE

Q. When did you decide to become a climate activist?

A. I can't pinpoint a specific time in my life. There was never a time when the climate crisis wasn't looming over my life and future. I was definitely concerned about climate change for as long as I can remember.

Q. How do you launch a climate organization?

A. You only start something new if you see there's a gap or there's something missing. Then you come up with a vision of what you want to fill that gap with. You communicate that with as many people as possible. You build a team around you and you just learn everything you need to as you go along. It's going to be really confusing, but you just keep going.

Q. If you could speak with the leaders of the world, what would you tell them?

A. If you can't look your child in the eyes and say you've done everything that you can to save their lives and future, it means you're not doing enough for them.

Q. What are your tips for kids who want to become climate activists?

A. Maybe you're an artist, maybe you like to dance, maybe you're good at computer programming or robotics or technology—whatever it is you're passionate about, use that to communicate the world you want to build and combat the climate crisis. I'm a writer and I'm a good speaker. I use my public speaking and my writing abilities to spread the message.

ISRA HIRSI

When Isra Hirsi joined an environmental club in her freshman year of high school, she didn't know it was going to be life changing. There, Isra, who's from Minneapolis, Minnesota, learned about climate change and decided to do something about it. Last year, she helped found an organization called U.S. Youth Climate Strike.

The organization helps organize climate strikes across the U.S. and encourages politicians to pass laws to cut greenhouse gas emissions. Isra also focuses on making the youth climate movement more inclusive. People of color and people who live in poverty are the most affected by climate change. For example, people who don't own an air conditioner are more likely to get sick during heat waves.

Isra wants people of color to have a voice. "Climate justice means seeing all voices incorporated in the fight against the climate crisis," she says.

Year born: 2003
From: Minneapolis, Minnesota
Favorite subject: History
Favorite outdoor activity: Biking
Favorite thing about planet Earth: Its colors
Favorite food: Spaghetti
Role model: Angela Davis (activist and author)
Dream job: Lawyer
Tip for aspiring climate activists: "It's okay to start small. The action you're doing in your community is amazing and is impacting people's lives. We need people like you to be a part of the movement."

QUICK Q&A WITH ISRA

Q. How is climate change affecting you personally?

A. I have family in Somalia, where there are droughts that have displaced millions of people. These droughts are causing starvation and food shortages.

Q. Why is it important for a woman of color like you to be a leader of the youth climate movement?

A. It's really important to have representation. Being a black woman and leading in this space automatically makes this space different. It leads into conversations that create real change not just for white voices but for everybody.

Q. What do you hope the youth climate movement will achieve?

A. I'm really hoping that, especially in the U.S., we'll help people realize how big this crisis is. On top of that, I hope politicians will listen to what young people are saying and take action.

Q. What are your tips for kids who want to become climate activists?

A. Talk to folks in your community about how the climate crisis impacts them. Make sure your own family is aware. Try to educate yourself, but also create awareness. Starting a school club or talking about it in your friend group is one way that kids can get involved.

JEAN HINCHLIFFE

In 2018, scientists at the United Nations released a report warning that the world has twelve years left to avoid the worst effects of climate change. The scientists said that to keep global warming in check, countries around the world have to cut their emissions of greenhouse gases by about half by 2030.

The report was a wake-up call for Jean Hinchliffe from Sydney, Australia. Jean grew up worrying about how climate change was affecting the planet. After reading about the report, she joined an organization called School Strike 4 Climate. The group helps organize climate strikes across Australia.

Jean says that being an activist and helping organize big rallies can be "chaotic." But the reward is worth it. "It's so exciting to see it all come together," she says. "It's also just a proud feeling knowing that you helped bring all of this together."

Year born: 2004
From: Sydney, Australia
Favorite subject: Art
Favorite outdoor activity: Roller-skating
Favorite thing about planet Earth: Its diversity of animals and plants
Favorite food: Raspberries
Role model: Greta Thunberg
Dream job: Activist and actress
Tip for aspiring climate activists: "Don't be afraid to ask for help."

QUICK Q&A WITH JEAN

Q. What about the United Nations report encouraged you to become a climate activist?

A. The fact that it gave a deadline of twelve years really scared me. I was fourteen years old when the report came out. I would be twenty-six by that deadline. It was just so scary that my politicians didn't care enough about my future to do anything about this.

Q. What do you do to address climate change in your personal life?

A. I've gone vegan. I also take public transport as much as I can, or I walk. I try to compost my food, as well as using reusable containers to generate less waste. But I think it's important to remember that while individuals can make their impacts, climate change is happening on a very big scale. What we want is a [systemic] change from burning coal and other sources of fossil fuels to renewable energy and a greener economy.

Q. What are your tips for kids who want to become climate activists?

A. You don't have to be organizing a giant rally to make a difference. You can start an environmental club in your school, you can start composting, or join a community group. Find a way that you can personally take action and, over time, see what else you can do.

Q. What's your life like as a climate activist?

A. Organizing these strikes is a very big task. We have lots of people organizing all across Australia as well as internationally. We are all constantly communicating online, as well as having weekly video meetings. At the local level, we have lots of different ways of organizing, from putting up posters all over a city to communicating with businesses to see if they will endorse us.

AYAKHA MELITHAFA

In 2018, the city of Cape Town, South Africa, had a serious problem: It was running out of water. Three years of drought had parched reservoirs for the city of four million people. Residents had to line up to collect water in tanks. Ayakha Melithafa experienced the crisis firsthand. Her mother is a farmer and some of her livestock died because of the water shortage. "That had a negative impact on our monthly income," Ayakha says.

The drought encouraged Ayakha to become a climate activist. She goes to climate strikes. In 2019, she also joined fifteen other young people, including Greta Thunberg and Alexandria Villaseñor, to file a complaint with the United Nations Committee on the Rights of the Child. The petition argues that several countries are violating children's rights by not addressing climate change.

Ayakha hopes that the petition will raise awareness about the climate crisis. "I hope that it can push our governments to act as soon as possible," she says.

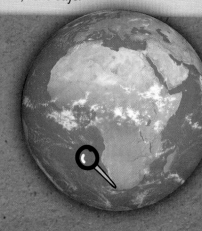

Year born: 2002
From: Cape Town, South Africa
Favorite subject: Life sciences
Favorite outdoor activity: Hiking
Favorite thing about planet Earth:
Its diversity of plants
and animals
Favorite food: Umphokoqo (South
African dish made with maize flour)
Role model: Her mom
Dream job: Environmental lawyer or engineer
Tip for aspiring climate activists: "No one is too little to make
a difference. Become who you want to be today, and raise
awareness about climate inaction so we can put enough
pressure on governments to make change."

QUICK Q&A WITH AYAKHA

Q. Why did you decide to join the petition with the United Nations?

A. South Africa has been really affected by climate change. It has droughts, flooding, and heat waves. Most unprivileged people and people who don't know much about climate change have to live with the consequences of climate change every single day. So I thought it wasn't fair that no one from South Africa was in this legal complaint. I felt like I had to represent my country.

Q. How does it make you feel to go to the climate strikes?

A. It feels amazing. It feels like I'm actually doing something that can impact our futures and bring change. But after striking, some people just go back to their normal lives. It's not just about striking. It has to be more than that. You have to be a climate-conscious person.

Q. What do you do to address climate change in your personal life?

A. With the climate organization Project 90 by 2030, I made my own solar-powered power bank to charge my phone. Also, I take shorter showers. I recycle more. I compost. I stopped asking for plastic bags. I use my own tote bags. And I stopped using plastic straws.

Q. What are your tips for kids who want to become climate activists?

A. You can start right now. You can start at home by taking shorter showers and encouraging your family to do the same. You can also raise more awareness about climate change. You can maybe start your own club at school. You can join activities like community cleanups.

35

RIDHIMA PANDEY

In 2013, several days of heavy rain caused destructive flooding and landslides in the state of Uttarakhand, in northern India. Thousands of people lost their lives. The disaster shocked Ridhima Pandey. She wasn't affected by the floods directly, but she lives in the same state. Ridhima asked her dad what they could do to help and so they got to work.

In 2017, she filed a petition against the Indian government for failing to address climate change and protect its citizens. In 2019, Ridhima also joined Ayakha Melithafa and other young people in their petition with the United Nations Committee on the Rights of the Child.

Ridhima hopes that by going to climate strikes and filing her petitions, her voice will be heard. "If children can understand the importance and the effects of climate change, why can't old people?" she says.

Year born: 2007
From: Haridwar, India
Favorite subjects: Environmental science and math
Favorite outdoor activity: Jungle safari
Favorite thing about planet Earth: The trees and animals
Favorite food: Her mom's food
Role model: A. P. J. Abdul Kalam (scientist and former president of India)
Dream job: Founder of a nonprofit organization to save the environment
Tip for aspiring climate activists: "Don't be afraid of anyone, and just raise your voice."

QUICK Q&A WITH RIDHIMA

Q. Why did you decide to join the petition with the United Nations?

A. It was a very good chance to take this issue globally because the whole world is going to face climate change. It's a good chance to secure our future.

Q. How does it make you feel to go to the climate strikes?

A. I feel really happy when I take part in all those climate strikes and I meet children like me. I'm so happy that so many children are aware of the climate crisis and they're making more children aware.

Q. If you could speak with the leaders of the world, what would you tell them?

A. I would tell them talk less and do more. I want them to do work instead of making statements. Please don't ignore us. Take this issue seriously because it's not a joke, it's real.

Q. What are your tips for kids who want to become climate activists?

A. Protest and make petitions, do whatever you want, but do the things that are right. Don't do anything illegal.

BE A CHANGE-MAKER

Feeling inspired already? You can be a climate hero, too! Though governments around the world can address climate change on a much larger scale, every one of us can make a difference. We all have a **carbon footprint**: We release CO_2 into the air because we need food to eat, clothes to stay warm, and transportation to move around. Ready to help fight climate change? If you have the means, here are some simple steps you can take to reduce your carbon footprint.

DID YOU KNOW?

A single person in the U.S. produces over 18 tons of CO_2 a year. That's about as heavy at 72,000 rolls of toilet paper!

SAY NO TO PLASTIC

Look around you: How many things can you count that are made of plastic? Plastic is a cheap and long-lasting material. So it's everywhere! But plastic is made with fossil fuels. The process of making plastic, and then disposing of it, produces emissions of CO_2. You can easily avoid single-use items like plastic water bottles and plastic bags. Drink water out of a reusable bottle that you can fill up. Bring your own tote bags when you go shopping.

Bonus: Avoiding plastic reduces pollution in the ocean. Animals there can die if they become entangled in plastic items or mistake plastic for food.

TURN OFF THE LIGHTS

Whenever you leave a room, remember to turn off the lights! You should also turn off your TV, computer, and other electronics when you don't need them. Or even better: Just unplug them. Many electronic devices keep sucking energy even when they're off! The energy powering your home likely comes from a fossil fuel source like oil or natural gas.

Bonus: Turning off unnecessary lights reduces light pollution, unwanted and overly bright light that harms wildlife like birds, bats, and frogs.

WALK OR BIKE

We all need to go places! If possible, try to get there by walking, biking, or taking public transportation instead of using a car. Otherwise, try to carpool and share a ride with friends—only if it's safe to do so! Cars, SUVs, and other personal vehicles produce one-fifth of all greenhouse gas emissions in the U.S.

Bonus: Walking and biking are a form of exercise and will help you stay healthy. Carpooling can be an opportunity to spend time with your friends.

GIVE FOOD SOME THOUGHT

Try to buy food that is produced locally as much as you can. Transporting items around the globe in ships and airplanes produces lots of greenhouse gas emissions. Labels can help you figure out where meat and produce come from. If you can, also try to reduce the amount of meat, especially beef, you eat. Raising livestock takes up lots of water and land that could otherwise be covered in trees. But remember: If you limit meat, you should still eat plenty of protein, iron, and vitamins by consuming other foods, like beans, nuts, and eggs.

Bonus: Eating lots of fruits and veggies is good for your health.

PLANT TREES

A great way to help fight climate change is by planting more trees. Plants remove CO_2 from the air and use it to grow. Talk to your family about planting seedlings in your backyard! If you don't have an outdoor space at your house, you can look for tree-planting events at your local park, botanical garden, or other community space.

Bonus: Trees create homes for all types of animals. Trees also create shade. That helps keep temperatures down, especially in cities.

TAKE SHORTER SHOWERS

5 min

The average person in the U.S. uses 80 to 100 gallons (303 to 379 liters) of water per day. A 10-minute shower can consume as much as 50 gallons (189 liters) of water! Keep your showers below five minutes to waste less water. Climate change increases the risk of droughts. So saving water whenever you can is important.

Bonus: Heating up water for your shower takes energy. So taking shorter showers also reduces greenhouse gas emissions.

REDUCE, REUSE, RECYCLE

You might have heard this motto before! It means that we should try to reuse things we already have, reduce the number of things we buy, and recycle materials whenever possible. That's because factories that make products use large amounts of fossil fuels. Try to reuse a plastic bag as many times as you can before throwing it away. Mend clothes instead of buying new ones, or organize clothing swaps with your friends when you feel the urge to get something new. Recycle paper, cans, and water bottles at your local recycling center.

Bonus: Reducing, reusing, and recycling keeps trash out of landfills.

STAY INFORMED

Scientists all over the world are studying all aspects of climate change to learn more about how it will affect us. Keep up with what's going on in your community and read the news to stay informed. Remember to always check the source and make sure it's well-known and trustworthy.

Bonus: When the public reads the news and is informed, our democracy is stronger.

THE CLIMATE ACTIVIST'S HANDBOOK

Lifestyle changes can help fight climate change. But in order to reduce greenhouse gas emissions on a large scale, governments around the world have to step in. It's the job of climate activists to try to sway public opinion and encourage world leaders to take action. Here are four steps you can take to become a budding climate activist!

Talk to Your Friends and Family

You care about climate change, but maybe the people around you don't! Talk to them about what climate change is and how it affects your community. Tell them how climate change makes you feel. Discussing your concerns with family and friends can help you feel better. It might also help them if they're feeling overwhelmed.

Make Your Own Group

If you can't find a local climate organization you like, or you think your community or school needs a different organization, launch your own! Talk to your friends about the ideas you have. Would they join your group? Ask a family member or a teacher you trust for help setting up a school group. Creating your own climate group will make you feel empowered.

Join a March

There's strength in numbers! Attending a climate strike or march with an adult in your family is a great way to make sure your voice is heard. Peaceful demonstrations can help raise awareness and encourage politicians to take action on a specific

issue. Create a funny or colorful sign to catch people's attention and express how you feel. Plus, attending a demonstration might help you make new friends who share your same interests!

Reach Out to Politicians

Politicians have the power to make laws that can address the climate crisis. Writing a letter to your local representative is a great way to let them know that you care about climate change and want them to take action. You can use the sample letter on page 44 as a template. Some climate organizations also organize events at which climate activists meet with politicians. That's an opportunity to speak with lawmakers face-to-face!

SPEAK UP

Use this template to write a letter to your state representative, senator, governor, or other local politician, to let them know you want them to take action on climate change. Ask a trusted adult for help mailing the letter.

Date:_____

Dear _____,

My name is _____ .

Add one more line to introduce yourself: _____

_____ .

Include a personal story that explains why you care about climate change: _____

_____

_____ .

Tell them what action you want them to take on climate change, like supporting renewable energy or climate change legislation, and why: _____

_____

_____ .

Write a conclusion that reinforces your major points:_____

_____

_____ .

Sincerely,
Your Name: _____

QUICK QUIZ

**Wondering what type of climate activist you can be?
Take the quiz to find out!**

1. **What part of this book did you find the most interesting?**
   a. Learning how I can address climate change in my personal life.
   b. Learning about different climate activists around the world and what they're doing to help.
   c. Learning about the science of climate change and its solutions.

2. **Do you know how climate change is affecting your community?**
   a. No, but I want to learn more.
   b. Yes, but I have more to learn.
   c. Of course! I'm a pro at this.

3. **In the next month, I will:**
   a. Use less plastic to limit my greenhouse gas emissions.
   b. Attend a climate strike or demonstration.
   c. Write to my representative to demand action on climate change.

4. **Why do you want to be an activist?**
   a. Changing my lifestyle to reduce my carbon footprint will make me feel empowered.
   b. I want all young people to come together and demand action on climate change.
   c. I want to encourage world leaders to take action.

5. **You're trying to convince a friend to join you in the fight against climate change. What are you most likely to tell your friend?**
   a. You can take very simple steps to address climate change, like walking more and taking shorter showers.
   b. Come with me to the next climate march! It'll be fun— you'll see!
   c. Make your voice count! Let's go to the city council or state legislature and tell politicians our future is at stake.

QUICK QUIZ RESULTS

Find out what type of climate activist you are by matching your answers with the results!

Mostly A's:

You believe in the power of every single person to make a difference. Go to page 38 to read more about how you can reduce your carbon footprint and help the environment.

Mostly B's:

You're ready to take to the streets! If you haven't done so already, join a trusted local or national climate movement with an adult in your family.

Mostly C's:

You believe people in power should solve the climate crisis. Let them know how you feel!

RESOURCES

Want to learn more about climate change and the youth climate movement? Check out these resources with permission and supervision of a trusted adult.

 Learn more about Earth's climate and weather: **https://climatekids.nasa.gov/**

 View how coastal areas could be affected by rising seas: **https://coast.noaa.gov/slr/**

 Watch videos, play games, and learn more about the oceans and our planet: **https://oceanservice.noaa.gov/kids/**

 For a fun dive into the science of climate change, read the book *The Magic School Bus and the Climate Challenge* by Joanna Cole and Bruce Degen.

 Read a collection of Greta Thunberg's speeches on climate change in the book *No One Is Too Small to Make a Difference* by Greta Thunberg.

 Read *The Tantrum That Saved the World* by Megan Herbert and Michael E. Mann, a fictional book about a young girl who rallies people to save the world from climate change.

GLOSSARY

activist: a person who actively advocates for political or social change

atmosphere: the protective layer of gases surrounding a planet

blizzard: a snowstorm with powerful winds that lasts longer than three hours

carbon footprint: the amount of greenhouse gas emissions produced by the activity of a person or group

climate change: a significant, long-term shift in weather patterns associated with a change in global temperatures

drought: a long period of little or no rain that has a negative impact on the water supply and/or crops

evaporate: to change from a liquid to a gas, or water vapor

fossil fuel: an energy source, such as coal, oil, or gas, formed inside the Earth from plant or animal remains

glacier: a large body of ice that moves slowly over land

greenhouse gas: any gas, such as carbon dioxide or methane, that traps heat in the atmosphere and causes Earth to warm up

heat wave: a period of abnormally high temperatures that lasts for more than five days

hurricane: a large storm with rotating winds of more than 74 miles (119 kilometers) per hour

meteorologist: a scientist who studies Earth's atmosphere and the weather

molecules: the smallest particles of chemical compounds, made up of two or more atoms bonded together

renewable energy: energy derived from sources such as the sun, wind, or water that are quickly replenished

vegan: a person who does not eat or use any animal products, including meat, dairy, eggs, and honey

wildfire: a fire that usually starts in the wild, such as a forest, and may spread to populated areas